Developing Numeracy

MEASURES, SHAPE AND SPACE

ACTIVITIES FOR THE DAILY MATHS LESSON

year

R

Ann Montague-Smith

A & C BLACK

Contents

Reprinted 2002
Published 2001 by A & C Black Publishers Limited
37 Soho Square, London WID 3QZ
www.acblack.com

ISBN 0-7136-5875-4

Copyright text © Ann Montague-Smith, 2001
Copyright illustrations © Martin Pierce, 2001
Copyright cover illustration © Charlotte Hard, 2001
Editors: Lynne Williamson and Marie Lister

The author and publishers would like to thank the following for their advice in producing this series of books: Madeleine Madden; Corinne McCrum

A CIP catalogue record for this book is available from the British Library.

A & C Black uses paper produced with elemental chlorine-free pulp, harvested from managed sustainable forests.

Printed in Great Britain by Caligraving Ltd, Thetford, Norfolk.

Introduction

Developing Numeracy: Measures, Shape and Space is a series of seven photocopiable activity books designed to be used during the daily maths lesson. They focus on the fourth strand of the National Numeracy Strategy *Framework for teaching mathematics*. The activities are intended to be used in the time allocated to pupil activities; they aim to reinforce the knowledge, understanding and skills taught during the main part of the lesson and to provide practice and consolidation of the objectives contained in the framework document.

Year R supports the teaching of mathematics by providing a series of activities which develop essential skills in measuring and exploring pattern, shape and space. On the whole the activities are designed for children to work on independently, although due to the young age of the children, the teacher may need to read the instructions with the children and ensure that they understand the activity before they begin working on it.

Year R teaches children to:

- use and understand the language of measure, pattern, shape, space and time;
- compare and order measurements of length, mass and capacity;
- understand the passing of time and to begin to tell the time;
- recognise and repeat simple patterns;
- use a variety of shapes to make models, pictures and patterns, and describe them;
- put sets of objects in order of size;
- use everyday words to describe position, direction and movement.

Extension

Many of the activity sheets end with a challenge (**Now try this!**) which reinforces and extends the children's learning, and provides the teacher with the opportunity for assessment. Again, it may be necessary to read the instructions with the children before they begin the activity. For some of the challenges the children will need to record their answers on a separate piece of paper.

Organisation

Very little equipment is needed, but it will be useful to have available: coloured pencils, counters, scissors, glue, dice and solid shapes. You will need to provide a 'days of the week' chart for page 26, a one-minute sand timer for page 28, Plasticine for page 35,

interlocking cubes for page 41 and beads and laces for page 53.

If possible, children should have access to a balance/objects to balance and different containers for pouring and filling to give them practical experience of mass and capacity.

Some of the sheets require cutting and sticking; this can be done by the children or an adult as appropriate.

To help teachers to select appropriate learning experiences for the children, the activities are grouped into sections within each book. However, the activities are not expected to be used in that order unless otherwise stated. The sheets are intended to support, rather than direct, the teacher's planning.

Some activities can be made easier or more challenging by masking and substituting some of the numbers. You may wish to re-use some pages by copying them onto card and laminating them, or by enlarging them onto A3 paper.

Teachers' notes

Very brief notes are provided at the foot of each page giving ideas and suggestions for maximising the effectiveness of the activity sheets. These can be masked before copying.

Structure of the daily maths lesson

The recommended structure of the daily maths lesson for Key Stage 1 is as follows:

Start to lesson, oral work, mental calculation	5–10 minutes
Main teaching and pupil activities *(the activities in the **Developing Numeracy** books are designed to be carried out in the time allocated to pupil activities)*	about 30 minutes
Plenary *(whole-class review and consolidation)*	about 10 minutes

Whole-class warm-up activities

The following activities provide some practical ideas which can be used to introduce or reinforce the main teaching part of the lesson.

Comparing and ordering measures

Length
Choose two items from a selection (skipping ropes, ribbons, strips of paper, etc), and ask the children questions such as: *Which is longer/shorter/wider/narrower?* When the children are confident, introduce a third item. The children will find it helpful if one end of each item is placed in line so that a direct visual comparison can be made.

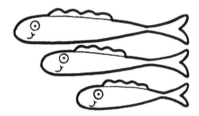

Mass
Choose two items from a selection (shoe, pencil case, wrapped parcel, etc.), and ask the children, sitting in a circle, to pass the items around for each child to hold briefly before deciding which is the heavier. Check with a bucket balance, and repeat.

Capacity
Provide a selection of containers (cups, spoons, a jug, a teapot, etc.), and colour some water with food colouring, so that its depth can be seen more easily. Explain that you will pour some water into a cup. Ask: *Which will hold more, the cup or the teapot?* Repeat for other pairs of containers.

Days of the week
Provide a date chart and ask the children to recite the days of the week in order. Ask: *What day is it today? What day will it be tomorrow? What day was it yesterday? Which days do we come to school? When is the weekend?*

Passing of time
During a PE lesson, show the children a one-minute sand timer and explain that you are going to see how many times they can, for example, run carefully around the room before the sand runs out. Compare results, and repeat for a different task.

Exploring pattern, shape and space

Children patterns
Arrange five girls and five boys in an alternating boy/girl/boy/girl pattern. Ask the children to describe the pattern and invite others to join the pattern. Repeat for different criteria, such as standing/sitting, colour of eyes, etc.

Behind the wall shapes
Put together a selection of 3-D shapes (cubes, cuboids, cylinders, pyramids, cones and spheres) **or** 2-D shape tiles (squares, rectangles, triangles, circles, stars, crescent moons). Gradually slide a shape up above a screen (piece of card or book) so that it just peeks over the top. Ask: *What shape might this be? Why do you think that?* Keep showing a little more of the shape until the children work out which shape it is and can name it correctly.

Feely bag shapes
Put a 3-D shape **or** a 2-D shape tile inside a feely bag and pass the bag around the group. Each child feels the shape, decides what it is, then passes the bag to the next child. When the bag has been all around the group, the children say the name of the shape. Ask questions such as: *How many sides/faces does the shape have? Is the shape flat or curved? Does it have straight or curved sides?*

PE positions, movements and directions
Ask the children to stand in different positions, (in front of, behind, next to, in between, opposite) a partner. Check that they understand the language and can follow the position instructions.

Ask the children to follow/give instructions such as: *walk forwards/backwards; turn left/right; face the front; go behind the bench.*

Turtle moves
Using programmable toys, such as Roamer or Pip, ask the children to suggest how to programme the toy to move it from one point to another. Encourage them to use the language of movement and direction, such as: *send it forward 5; turn it left 2.*

Bones

This bone is longer .

This bone is shorter .

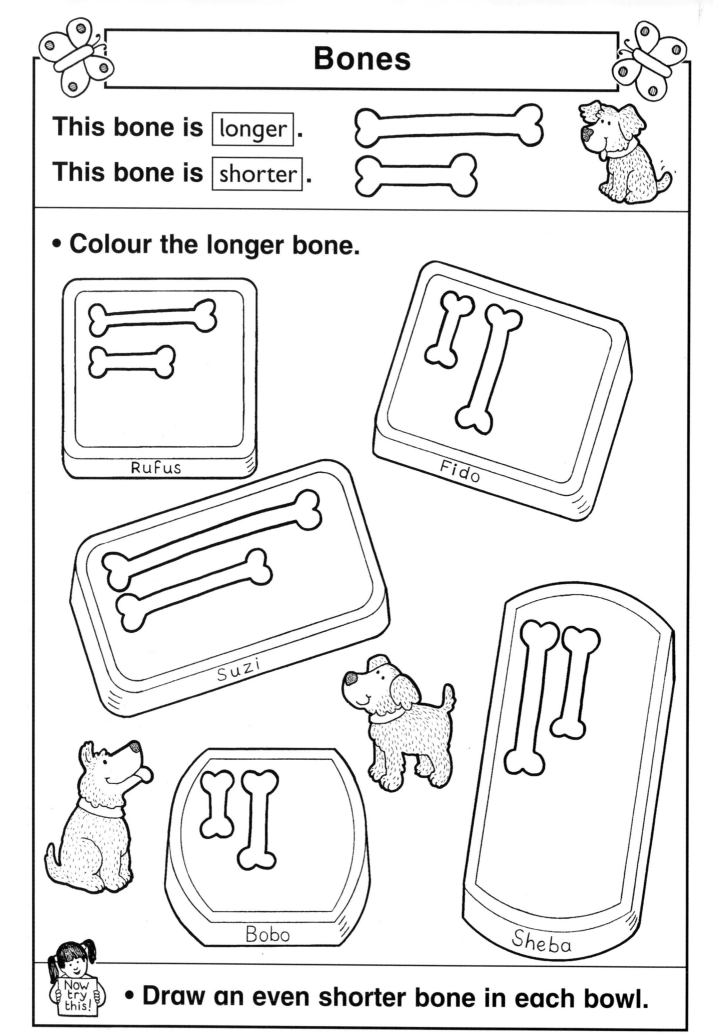

• **Colour the longer bone.**

Rufus

Fido

Suzi

Bobo

Sheba

• **Draw an even shorter bone in each bowl.**

Teachers' note Ensure that the children understand that when measuring and comparing two or more lengths, the lengths must be in alignment with each other at one end. Demonstrate this to the children by comparing, for example, two pencils.

Developing Numeracy
Measures, Shape and Space
Year R
© A & C Black

Ears

shorter ears   longer ears

• **Draw the missing ears.**

shorter   longer   shorter   longer

shorter   longer   shorter   longer

 • **Draw the missing ears.**

shorter   longer

Teachers' note Ask the children to find examples of shorter and longer objects, such as comparing two ribbons or two sticks for length. Encourage them to use the vocabulary of comparing length.

**Developing Numeracy
Measures, Shape and Space
Year R
© A & C Black**

7

The bathroom

- **Colour the** longest **blue.**
- **Colour the** shortest **red.**

longest

shortest

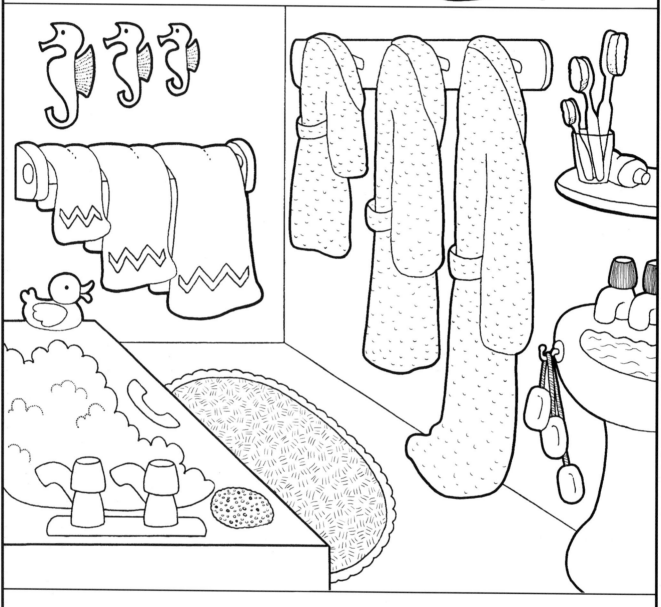

- **Cut out the toothbrush.**
- **Find something longer. Draw it.**
- **Find something shorter. Draw it.**

Teachers' note Ask the children to find the sets of three before they start the activity. Encourage the children to run their fingers along the pictures to determine which are longer/shorter.

**Developing Numeracy
Measures, Shape and Space
Year R
© A & C Black**

8

Shortest and tallest

- **Colour the** shortest **red.**
- **Colour the** tallest **blue.**

shortest   tallest

Teachers' note Encourage the children to make comparisons of the objects in each set by discussing which is the shortest and which the tallest. Then ask them to describe the size of the other objects in the set.

Developing Numeracy
Measures, Shape and Space
Year R
© A & C Black

Matching fish

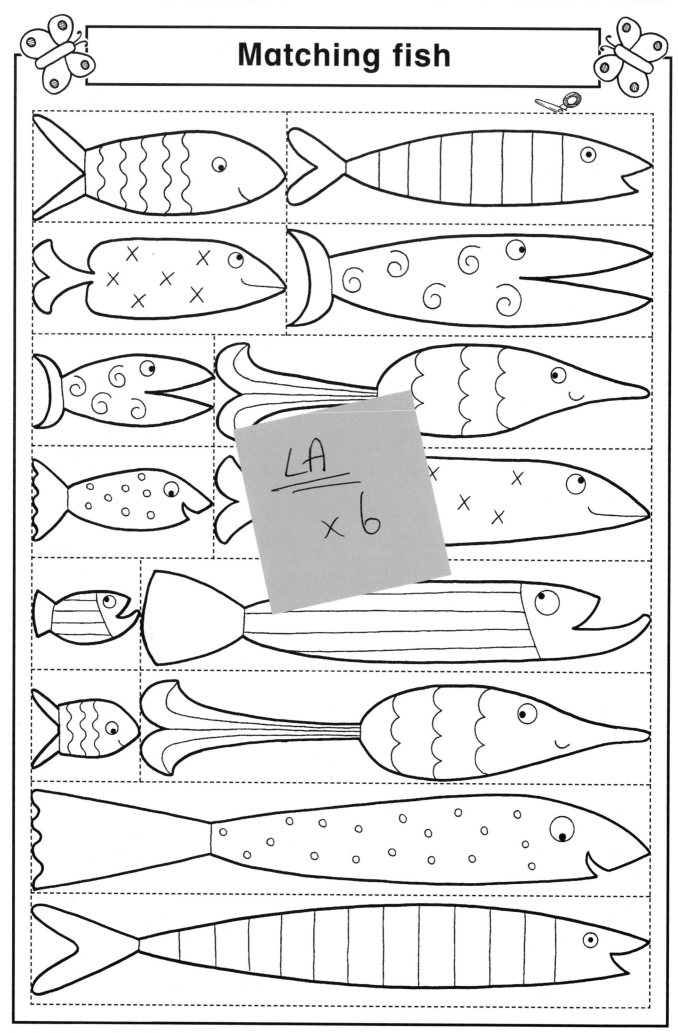

Teachers' note Cut out the cards and spread them face up. The children take turns to choose two fish which they think are the same length. If they are the same they keep them, if not, they put them back. These cards can also be used for ordering lengths and for playing Pelmanism by matching patterns.

**Developing Numeracy
Measures, Shape and Space
Year R
© A & C Black**

10

About the same

- **Colour the clothes that are** about the same **length.**

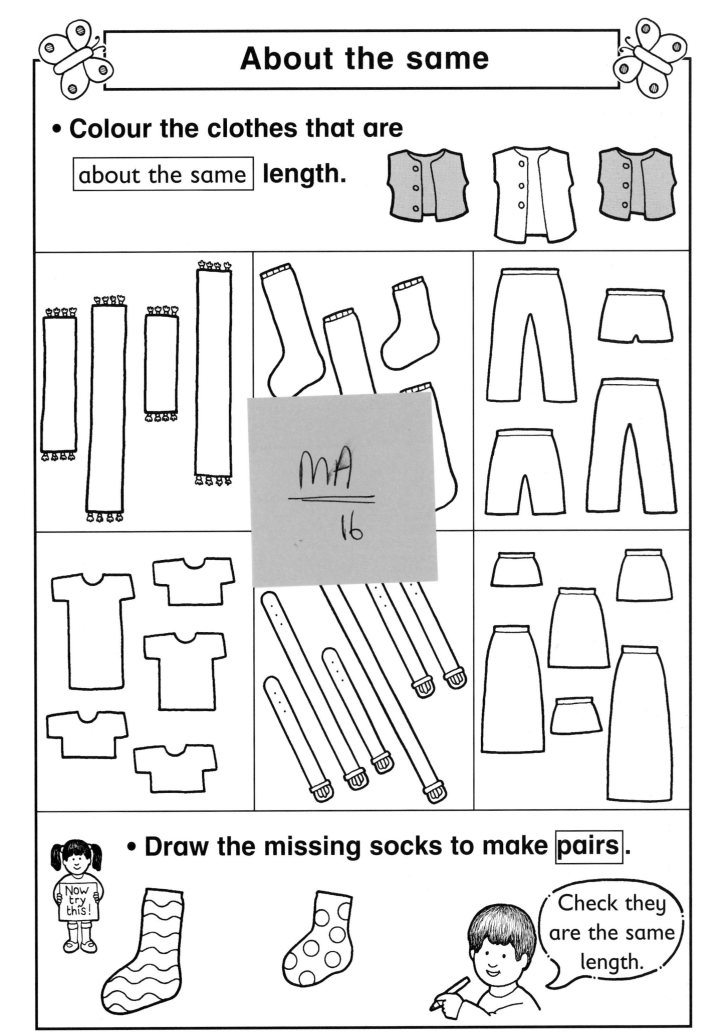

- **Draw the missing socks to make** pairs .

Now try this!

Check they are the same length.

Teachers' note Encourage the children to compare each item and to find all in each set that are about the same length. In the plenary, discuss with the children how they could check whether they were correct.

**Developing Numeracy
Measures, Shape and Space
Year R
© A & C Black**

11

Posting letters

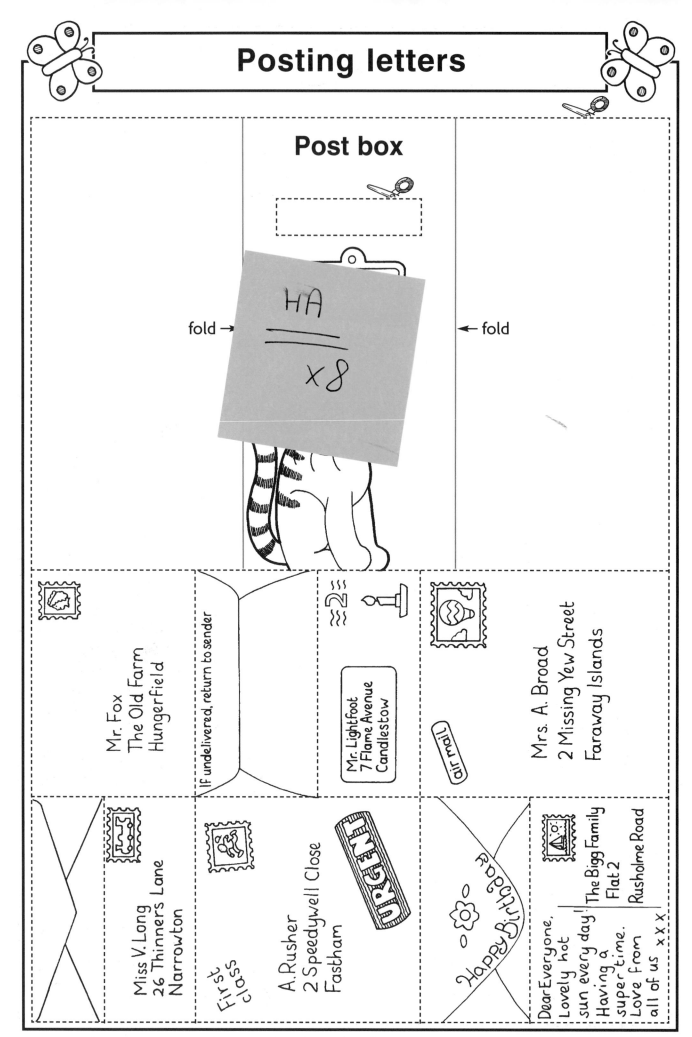

Post box

fold → ← fold

HA
―――
×8

Mr. Fox
The Old Farm
Hungerfield

If undelivered, return to sender

Mr. Lightfoot
7 Flame Avenue
Candlestow

air mail

Mrs. A. Broad
2 Missing Yew Street
Faraway Islands

Miss V. Long
26 Thinners Lane
Narrowton

First class

A. Rusher
2 Speedywell Close
Fastham

URGENT

Happy Birthday

Dear Everyone,
Lovely hot
sun every day!
Having a
super time.
Love from
all of us x x x

The Bigg Family
Flat 2
Rusholme Road

Teachers' note Copy the sheet onto card and cut along the dots to make a postbox and letters. Ask the children to put the letters in two piles: those that are narrow enough to fit through the post box and those that are too wide. Tell them to discuss their choices and to check by 'posting' the letters. As an extension, the children could draw and cut out letters to go through the post box.

Developing Numeracy
Measures, Shape and Space
Year R
© A & C Black

Weighing machines

Teachers' note Cut out the pictures. Working in groups, the children sort the pictures into two piles: those that show the scales being used correctly and those that show the scales being used incorrectly. During the plenary, discuss the children's choices. Ask: 'Which scales would you choose to weigh yourself?' 'Why?' 'What else could you weigh on the scales?'

**Developing Numeracy
Measures, Shape and Space
Year R
© A & C Black**

Does it balance?

- **Tick the ones that** balance .
- **Cross the ones that do not.**

Teachers' note Through practical experience of balancing, check that the children understand when objects do not balance. Encourage them to check that the pans are balanced before they begin weighing.

14

Developing Numeracy
Measures, Shape and Space
Year R
© A & C Black

Will they balance?

- ✓ **Tick the toys that will balance.**
- ✗ **Cross the toys that will not.**

- **Find two things that will balance.**
- **Draw them.**

Teachers' note Through practical experience of balancing, check that the children understand when objects do not balance. This page is designed to generate discussion, so the children should work in pairs or small groups. In the extension activity, the children should use a balance for checking.

Developing Numeracy
Measures, Shape and Space
Year R
© A & C Black

15

Heavier and lighter

The penguin is ⬚lighter⬚.

The lion is ⬚heavier⬚.

• ✓ **Tick the heavier animal.**

• ✓ **Tick the lighter animal.**

• **Draw a heavier animal.**

• **Draw a lighter animal.**

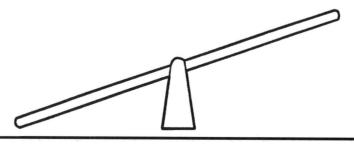

Teachers' note Talk about the children's experiences of using a see-saw. Discuss what the 'up' and 'down' positions of the see-saw mean. This page is designed to generate discussion, so the children should work in pairs or small groups.

**Developing Numeracy
Measures, Shape and Space
Year R
© A & C Black**

Vegetables

This carrot is ⎡lighter⎤.

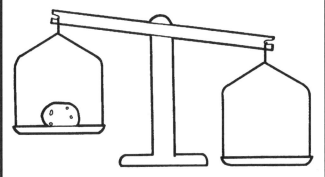

This carrot is ⎡heavier⎤.

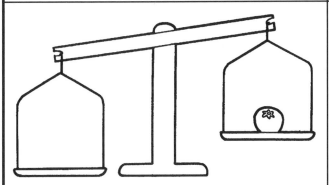

Draw a heavier potato.

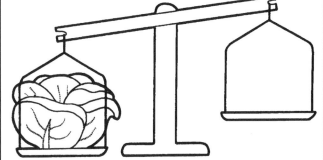

Draw a lighter cabbage.

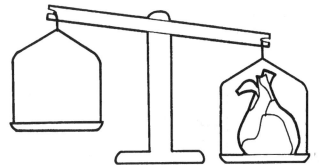

Draw a heavier tomato.

Draw a lighter onion.

- **Draw the heaviest vegetable you know.**

- **Draw the lightest vegetable you know.**

Teachers' note Demonstrate this activity practically and encourage the children to think about the size of the vegetables and the weight. Compare the weights of different vegetables. Is it always the biggest that is the heaviest? Begin to introduce the concept that size does not always signify the weight of an object.

Developing Numeracy
Measures, Shape and Space
Year R
© A & C Black

Transport

- **Colour the** heaviest **red.**
- **Colour the** lightest **blue.**

heaviest   lightest

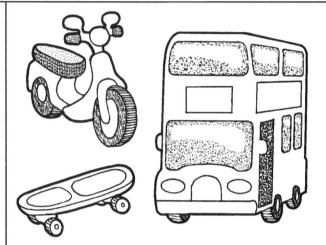

Now try this!

- **Draw something**

  heavier than a bus.

- **Draw something**

  lighter than a bus.

Teachers' note Discuss different modes of transport and ensure that the children realise that the pictures depict real objects, not toys. Encourage the children to talk about the objects in the pictures and to explain their choices for heaviest and lightest.

**Developing Numeracy
Measures, Shape and Space
Year R
© A & C Black**

Fill it up

Salt

Coughix
for children

soup

Teachers' note Provide the children with the opportunity to explore the relative capacities of containers using sand and water. Cut out the cards and ask the children to match the things that go together. Encourage them to recognise the features that make the pairs match. Can the children think of other 'pairs' of containers?

**Developing Numeracy
Measures, Shape and Space
Year R
© A & C Black**

19

The kitchen

- **Draw lines to show which holds more .**

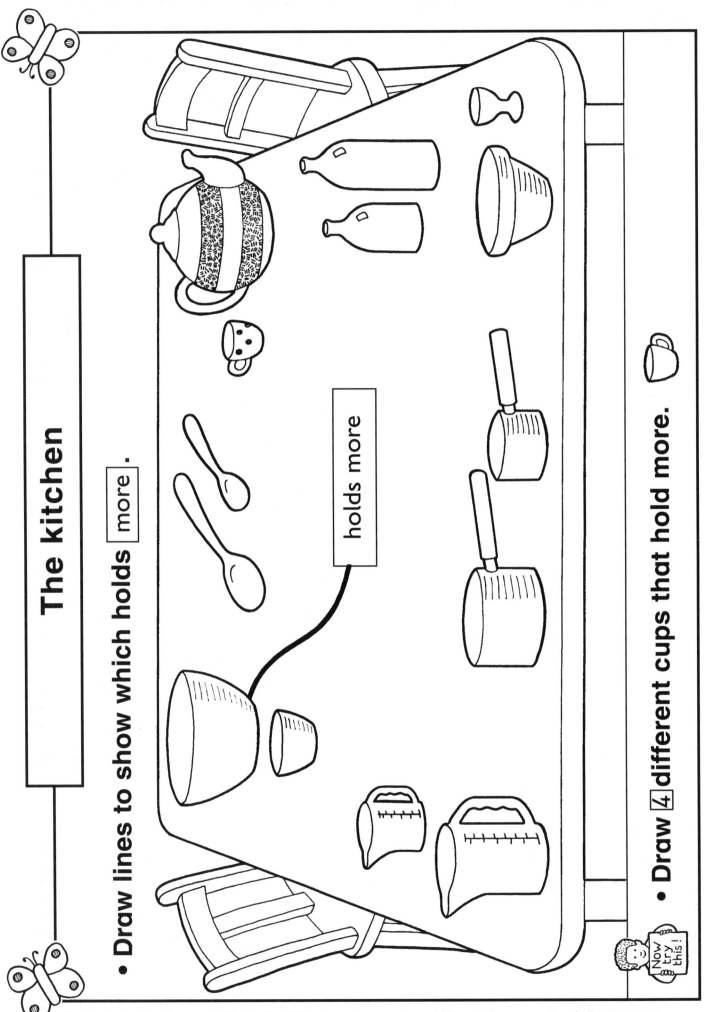

holds more

- **Draw 4 different cups that hold more.**

Teachers' note Provide opportunities for the children to explore pouring and filling similar containers, for example large and small water bottles. Discuss their observations, using the vocabulary 'more' and 'less'. When completing the activity, the children should look at each pair in turn and decide which one holds more.

20

Developing Numeracy
Measures, Shape and Space
Year R
© A & C Black

More and less

This vase has | more | **water.**    **This vase has** | less | **water.**

- **Draw the missing water.**

  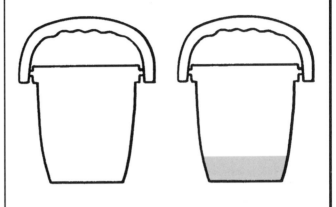

more   less   more   less

 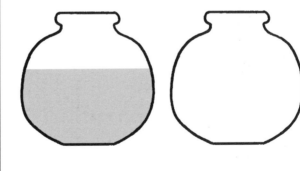

more   less   more   less

 • **Draw the missing water.**

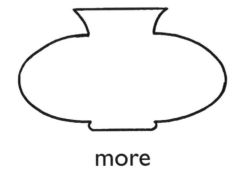 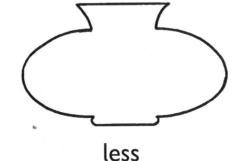

more   less

Teachers' note Demonstrate pouring water into two identical containers so that one holds more.
Encourage the children to describe what they see, using the language of 'more' and 'less'.

**Developing Numeracy
Measures, Shape and Space
Year R**
© A & C Black

Bubble bath

- **Look at each set of bottles.**

- **Colour blue the bottle holding the** [least] **.**

- **Look at each set of bottles.**

- **Colour red the bottle holding the** [most] **.**

Teachers' note Provide the children with experiences of filling and pouring. Ask them to describe how full the containers are using the language of capacity.

22

**Developing Numeracy
Measures, Shape and Space
Year R
© A & C Black**

Vase snap

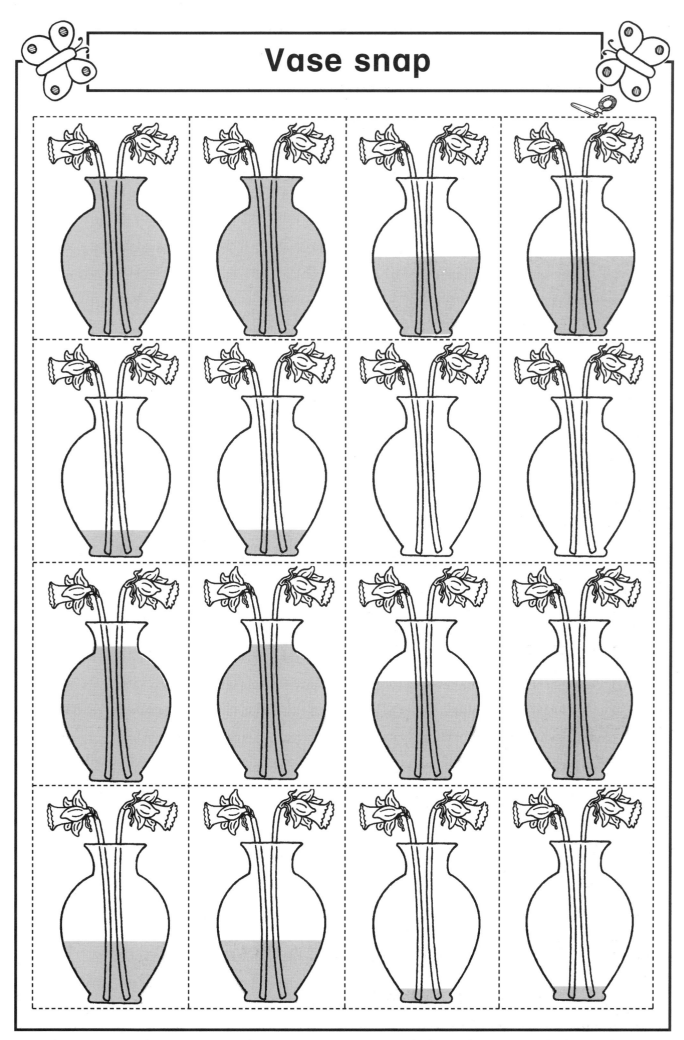

Teachers' note As an alternative to Snap, ask the children to play Pelmanism (see page 32). Encourage the children to describe how full each vase is using vocabulary such as: full, empty, half, almost, nearly.

**Developing Numeracy
Measures, Shape and Space
Year R
© A & C Black**

Story time

- **Colour to show**

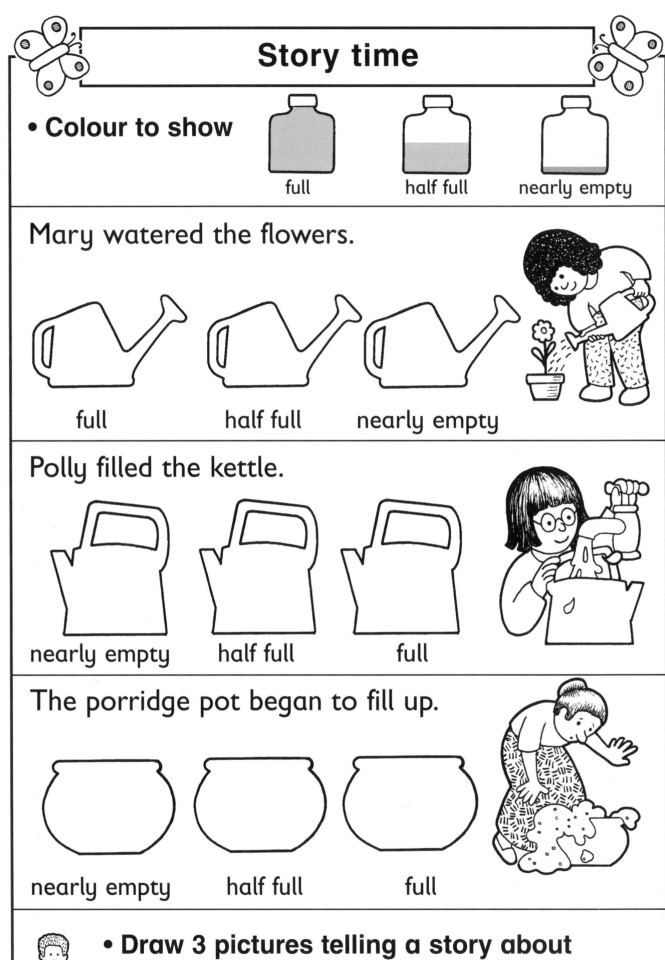

full   half full   nearly empty

Mary watered the flowers.

full   half full   nearly empty

Polly filled the kettle.

nearly empty   half full   full

The porridge pot began to fill up.

nearly empty   half full   full

- **Draw 3 pictures telling a story about a pot that filled up or emptied.**

Teachers' note Remind the children of the meaning of 'full', 'half full' and 'nearly empty'. Remember the stories/rhymes together.

Developing Numeracy
Measures, Shape and Space
Year R
© A & C Black

Rhyme time

Teachers' note Recite the nursery rhymes together to remind the children of the stories. Cut out the cards (four for each rhyme) and ask the children to put them in the correct order. As an extension, the children could draw four pictures to illustrate another rhyme and give them to a partner to reorder.

**Developing Numeracy
Measures, Shape and Space
Year R**
© A & C Black

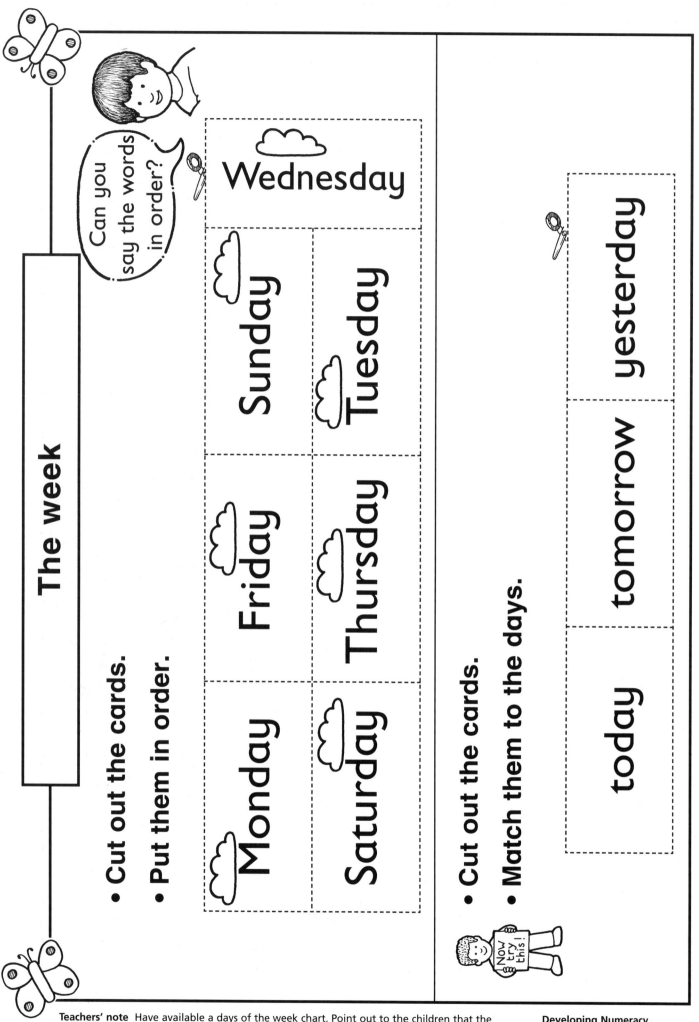

The week

Can you say the words in order?

- Cut out the cards.
- Put them in order.

Wednesday

Sunday

Tuesday

Monday

Friday

Thursday

Saturday

- Cut out the cards.
- Match them to the days.

yesterday

tomorrow

today

Now try this!

Teachers' note Have available a days of the week chart. Point out to the children that the position of the clouds will help them sequence the days. For the extension activity, the children could work in pairs. The sequenced days could be stuck onto a long strip of paper and joined to make a cylinder to reinforce the cyclical nature of the days of the week.

Developing Numeracy
Measures, Shape and Space
Year R
© A & C Black

My day

Teachers' note Cut up the cards for the children to sequence. As an extension, the children could draw the next stage, for example brushing teeth, reading a bedtime story. Discuss their day with the children. Ask for suggestions of what the children do when *they* get up, before *they* go to bed, and in the playground. After each suggestion, ask them what they do next.

**Developing Numeracy
Measures, Shape and Space
Year R**
© A & C Black

27

I minute

Use a I-minute sand timer.

- **Carefully colour in the apples.**

Stop when the sand has gone.

- **Count how many apples you coloured.**

I coloured ☐ apples.

Teachers' note Introduce the children to a one-minute sand timer. Explore how much can be achieved in one minute, for example how many bricks can be put together, how much of a nursery rhyme can be sung, how many names of children can be said.

Developing Numeracy
Measures, Shape and Space
Year R
© A & C Black

Telling the time

• **Join the pictures to the clocks.**

7 o'clock

9 o'clock

12 o'clock

5 o'clock

• **Draw yourself going to bed.**

I go to bed at ☐ o'clock.

Teachers' note Provide clocks so that the children can set the hands to the times shown. During the plenary, invite some children to demonstrate their bedtime on a clock. Ask: 'What time does Aaron go to bed?'

Developing Numeracy
Measures, Shape and Space
Year R
© A & C Black

Time snap

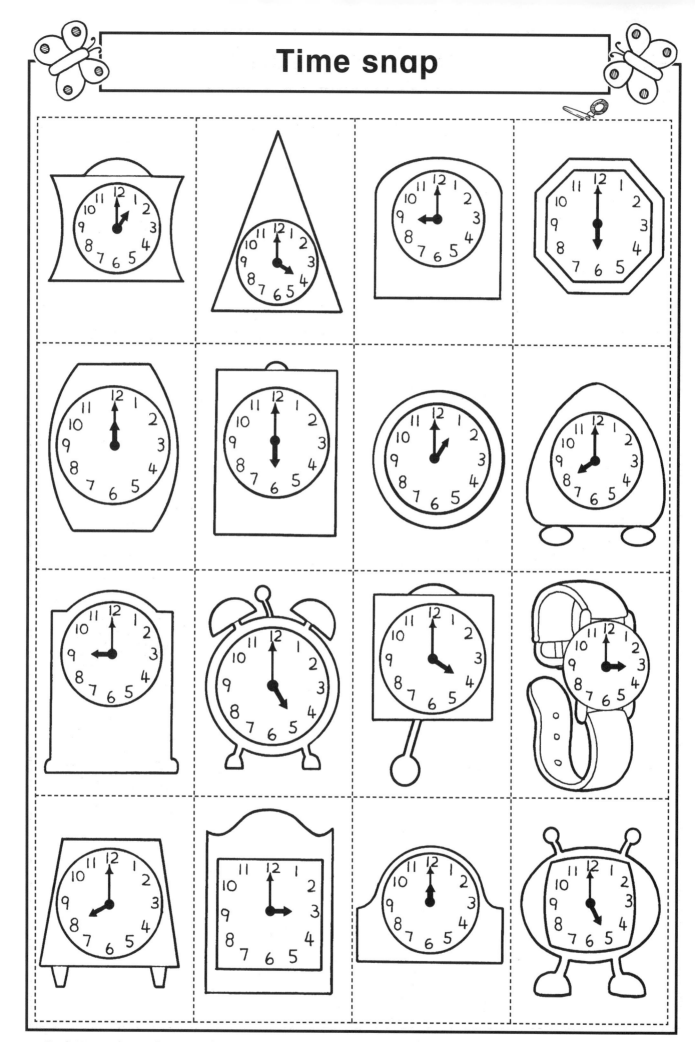

Teachers' note As an alternative to Snap, ask the children to play Pelmanism (see page 32). As an extension, provide the written time for each clock and ask the children to match the written time to the correct clock face.

**Developing Numeracy
Measures, Shape and Space
Year R**
© A & C Black

30

3-D shape match

- ## Colour each shape a different colour.

cube   sphere   pyramid   cone

- ## Colour these shapes to match the ones above.

Teachers' note Encourage the children to discuss the properties of the shapes: flat and curved faces, edges, corners. Some children may need to match the pictures to real solid shapes. As an extension, ask the children to identify and draw cubes and spheres they see around the room.

Developing Numeracy
Measures, Shape and Space
Year R
© A & C Black

3-D shape snap

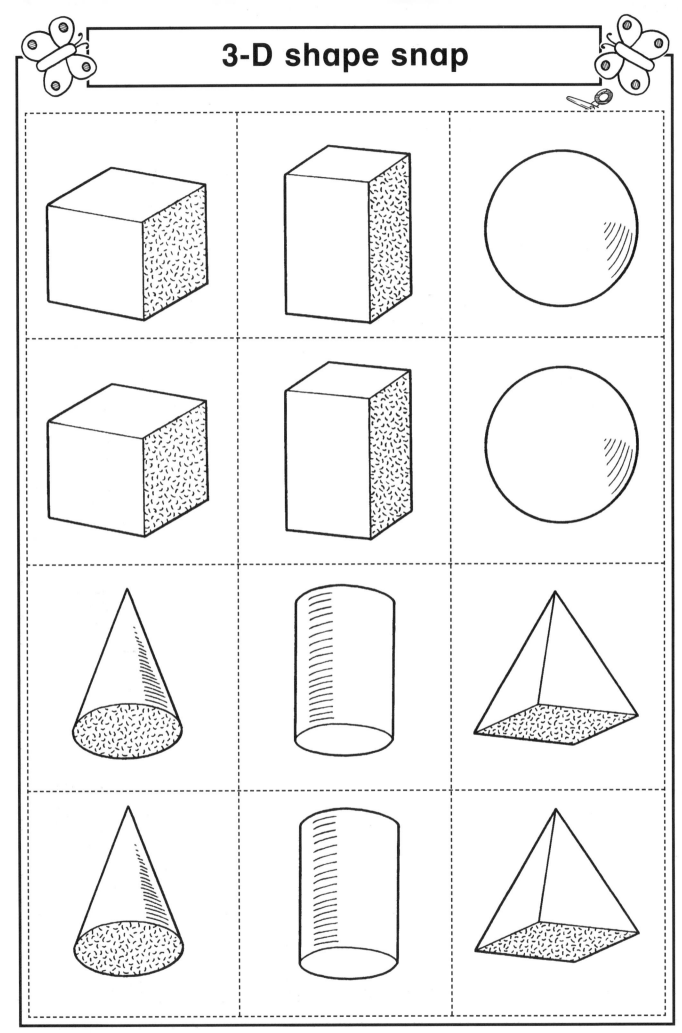

Teachers' note As an alternative to Snap, ask the children to play Pelmanism. Spread the cards face down. The children take turns to turn over two cards. If they match they keep them, if not, they turn them back. Encourage them to try to remember the position of the shapes they have seen. The winner is the child with the most cards at the end of the game.

**Developing Numeracy
Measures, Shape and Space
Year R**
© A & C Black

3-D shape sort

• Join matching shapes.

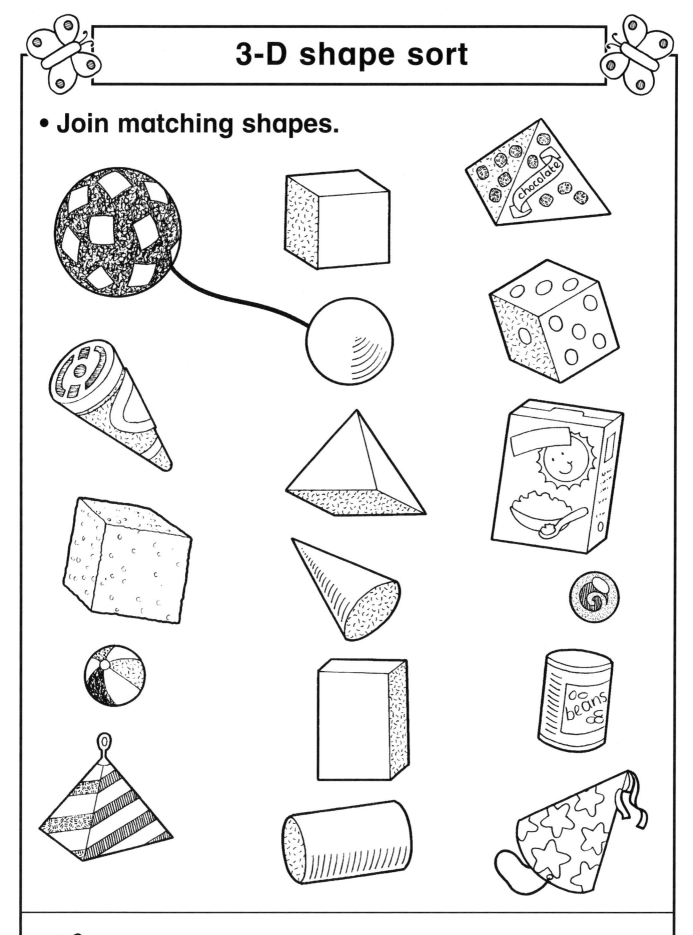

• Colour green the shapes that have a square `face`.

Teachers' note Encourage the children to name everyday objects by their mathematical shape, for example: 'This ball is a sphere'.

**Developing Numeracy
Measures, Shape and Space
Year R**
© A & C Black

Build a model

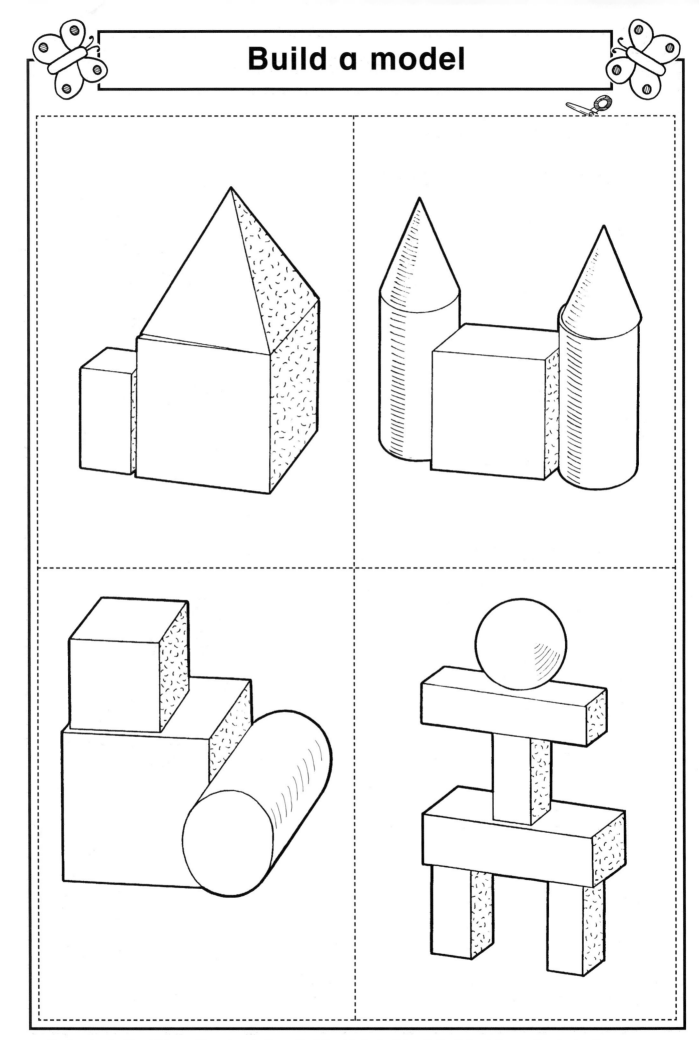

Teachers' note The children will need a selection of building blocks for this activity. Cut out the pictures and ask each group to copy one. When they have finished, swap the pictures around. In the plenary, discuss which shapes were easiest to build with, and why. As an extension, the children could make their own models from building blocks or junk material.

Developing Numeracy
Measures, Shape and Space
Year R
© A & C Black

34

Making shapes

- **Find a** ⬚sphere⬚ **like this.**

- **Copy it using plasticine.**

- **Now make the other shapes in your set.**

- **Make these spheres.**

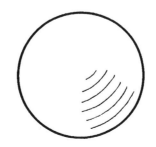

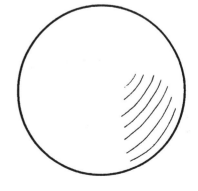

Teachers' note Provide the following solid shapes for the children to copy: sphere, cube, cuboid, cone, cylinder, square-based pyramid. Encourage the children to describe the shapes as they make them, i.e. curved/flat faces.

**Developing Numeracy
Measures, Shape and Space
Year R
© A & C Black**

2-D shape match

- **Join the dots.**
- **Match the shapes.**

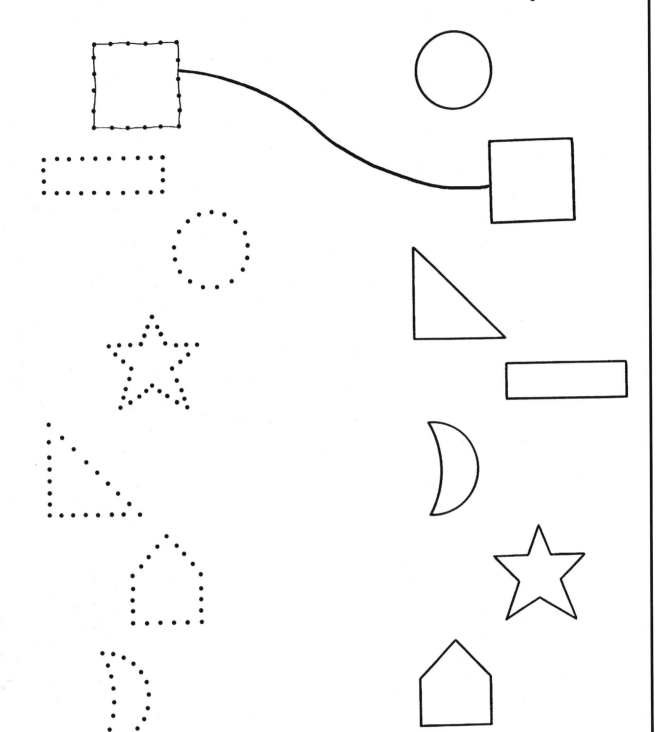

- **Colour shapes with** | straight edges | **red.**
- **Colour shapes with** | curved edges | **blue.**

Teachers' note Allow the children to touch and talk about the shapes before they complete this activity. Encourage them to name the shapes and to talk about their properties: number of sides; straight and curved sides.

**Developing Numeracy
Measures, Shape and Space
Year R
© A & C Black**

36

Shape town

- **Colour each shape a different colour.**

 square   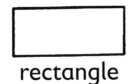 rectangle   triangle   circle

- **Colour the** faces **to match the shapes above.**

- **Find things in the classroom**

  with a face **this shape.**
- **Draw them.**

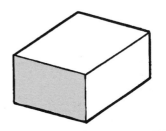

Teachers' note Discuss with the children the faces of shapes that they see every day, for example on a wall or a roof. Encourage them to connect the solid shape to its face, for example a cube having square faces. Take the children outside and look at the building. Discuss the shapes that they can see.

**Developing Numeracy
Measures, Shape and Space
Year R
© A & C Black**

Smiley faces

- **Colour the** face **marked** ☺ .
- **Colour the matching shape.**

	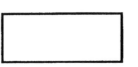
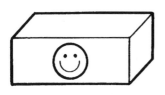	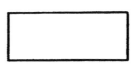
	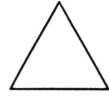
	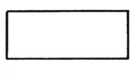 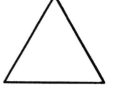
	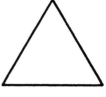

 • **Find things in the classroom**
  with a face **this shape.**
- **Draw them.**

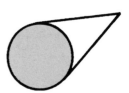

Teachers' note Encourage the children to examine real solid shapes and to compare their faces with the drawings.

**Developing Numeracy
Measures, Shape and Space
Year R
© A & C Black**

2-D shapes

Teachers' note Use this in conjunction with page 40. Cut out the shape cards and ask the children to sort them onto the machine on page 40.

**Developing Numeracy
Measures, Shape and Space
Year R
© A & C Black**

39

Sorting machine

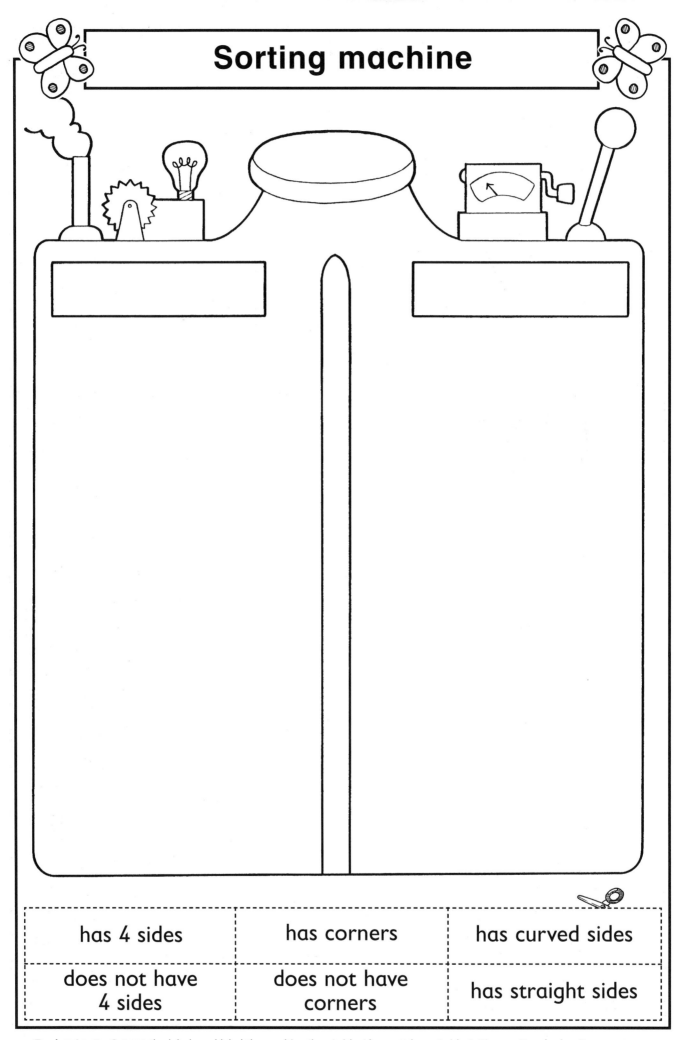

| has 4 sides | has corners | has curved sides |
| does not have 4 sides | does not have corners | has straight sides |

Teachers' note Cut out the labels and label the machine 'has 4 sides/does not have 4 sides'. Give the children the cards cut from page 39 and ask them, working in pairs, to sort the shapes onto the machine. Encourage them to discuss the properties of the shapes as they do so. As an extension, substitute the labels on the machine and ask the children to sort the shapes again.

Developing Numeracy
Measures, Shape and Space
Year R
© A & C Black

Circle beware!

start

finish

Teachers' note The children take turns to throw a die and move their counter. They count the sides of the shape they land on and take that number of interlocking cubes. If they land on a circle they lose a cube. The winner is the player with the longest rod of cubes. The game can be adapted for more able children – they gain an extra cube if they can say something else about the shape.

Developing Numeracy
Measures, Shape and Space
Year R
© A & C Black

41

Shape monsters

- **This monster is made from** squares .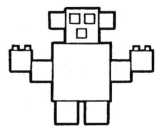

- **Make a monster from** circles .

- **Make a monster from** triangles .

 • **Make a monster from one of these shapes.**

Teachers' note Encourage the children to draw the shapes carefully so that their properties are clear. Some children may need to have the first shape drawn for them. As a further extension, the children could draw monsters using more than one type of shape.

Developing Numeracy Measures, Shape and Space Year R © A & C Black

Shape pictures

- **Cut out the shapes.**

- **Match the shapes to the pictures.**

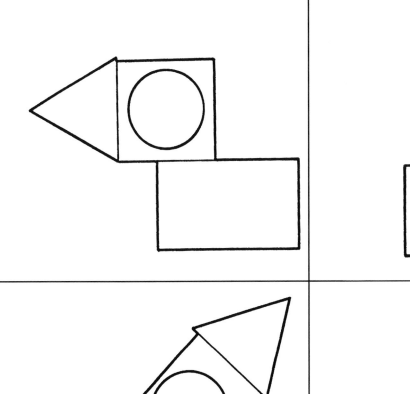

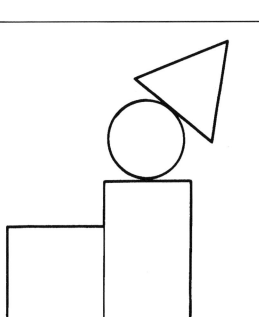

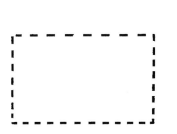

Teachers' note The children should match the shapes to the pictures, naming the shapes as they do so. As an extension, the children could use the shapes to make their own picture for a partner to copy.

**Developing Numeracy
Measures, Shape and Space
Year R**
© A & C Black

Pattern match

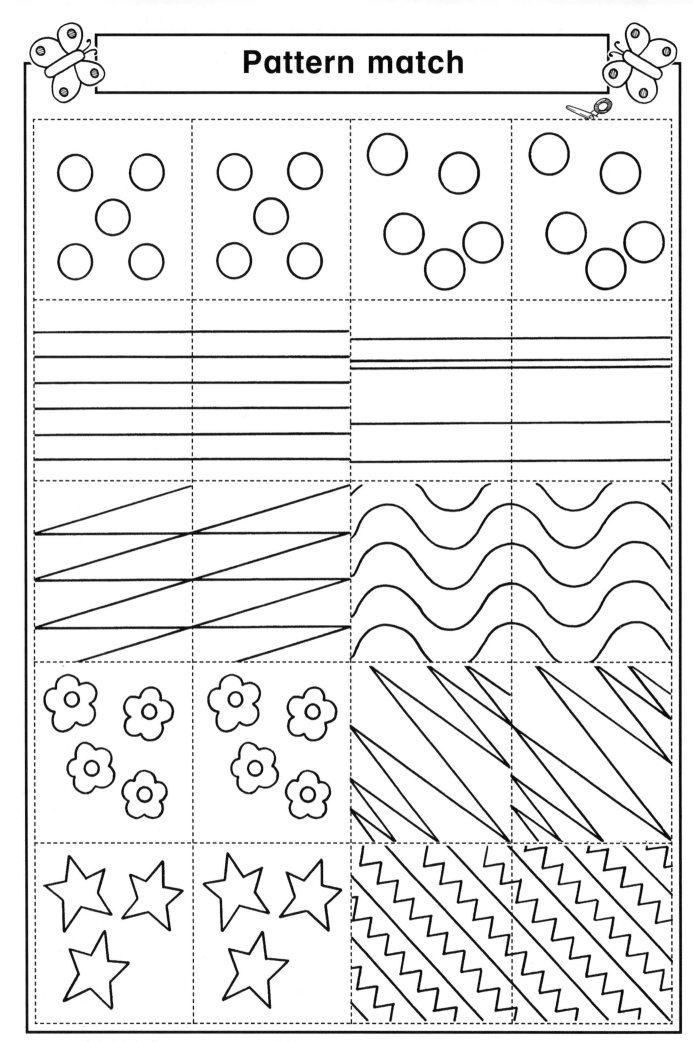

Teachers' note Cut out the cards and ask the children to match the patterns that go together. Encourage them to describe the shapes and patterns as they do so. The children can also use the cards to play Snap and Pelmanism (see page 32).

**Developing Numeracy
Measures, Shape and Space
Year R
© A & C Black**

Songs and rhymes

- **Cut out the cards.**
- **Put them in order.**

Start with the smallest.

Little Bo Peep has lost her sheep.

Girls and boys come out to play.

Five little speckled frogs.

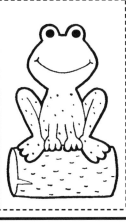    

Teachers' note Say or sing the rhymes together. Choose three children to stand at the front of the class and ask the other children to order them, smallest to largest. Encourage the children to use the language of ordering, for example 'between', 'next to'.

Developing Numeracy
Measures, Shape and Space
Year R
© A & C Black

Shape families: 1

• **Cut out the shape families.**

• **Draw your own shape family.**

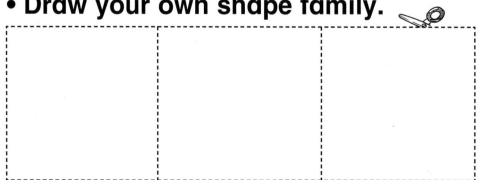

Teachers' note Use this in conjunction with page 47. As an extension, the children can draw their own shape family to order by size (smallest to largest, or vice versa) on page 47.

**Developing Numeracy
Measures, Shape and Space
Year R
© A & C Black**

Shape families: 2

● **Stick the shape families in order of size.**

◯			
△			
☐			
▭			

● **Stick your own shape family in order.**

Teachers' note Encourage the children to order each shape family (cut from page 46) in order of size. Explain that the first two families should be ordered from largest to smallest, and the second two families from smallest to largest.

Developing Numeracy
Measures, Shape and Space
Year R
© A & C Black

Monster families

• **Draw the missing monster.**

smallest middle-sized largest

smallest middle-sized largest

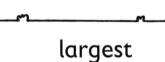

smallest middle-sized largest

• **Draw your own monster family.**

smallest middle-sized largest

Teachers' note Encourage the children to look carefully at the sizes of the monsters. Check that the monsters they draw are a suitable size.

**Developing Numeracy
Measures, Shape and Space
Year R
© A & C Black**

The Three Bears

• **Draw the missing things for each bear.**

• **Draw a bed for each bear.**

Teachers' note Remind the children of the story of *The Three Bears*. Ask them to draw the objects in the correct size order. Explain that the objects are the same shape but different sizes.

Developing Numeracy
Measures, Shape and Space
Year R
© A & C Black

49

Animal families

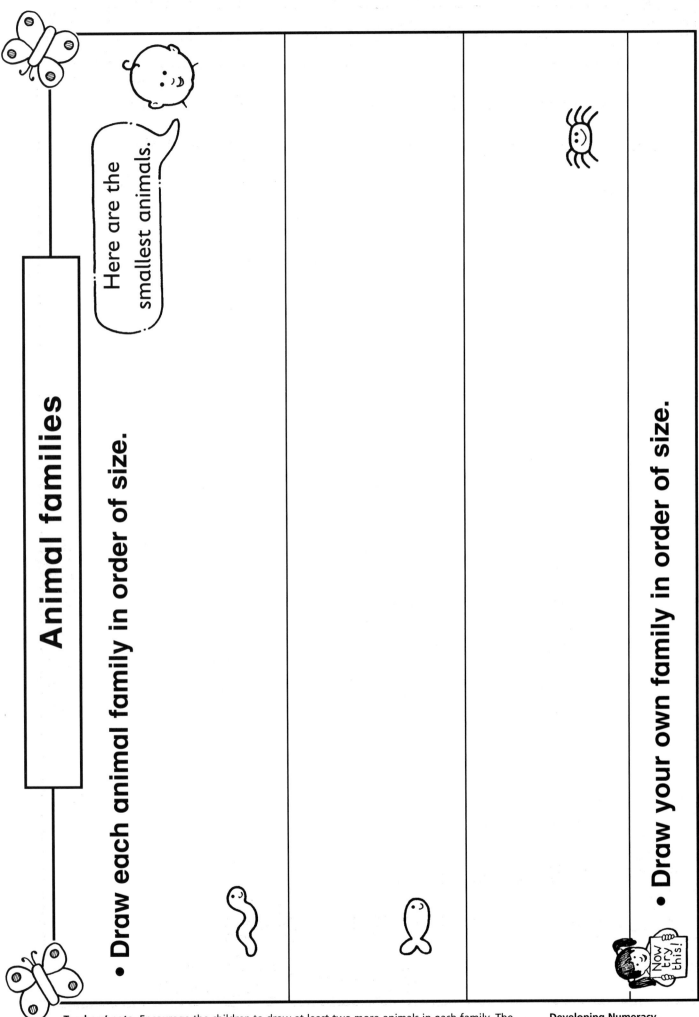

Here are the smallest animals.

• Draw each animal family in order of size.

• Draw your own family in order of size.

Teachers' note Encourage the children to draw at least two more animals in each family. The more confident children can draw three or four more animals, making each one slightly larger than the previous one. Note that the spider family is ordered from largest to smallest.

**Developing Numeracy
Measures, Shape and Space
Year R**
© A & C Black

Jack and the Beanstalk

Jack is planting the bean
seed next to the house.

• Draw the beanstalk.

It grew taller	and taller	and taller	and taller!

• **Draw a beanstalk as**

tall as yourself.

• **Work with a partner.**

Teachers' note Remind the children of the story of *Jack and the Beanstalk*. Encourage them to pretend that they are the beanstalk. Ask them to curl up, then gradually uncurl, growing taller and taller and taller. For the extension activity, the children may need rough paper, scissors and glue. Encourage them to think carefully about what they will need.

**Developing Numeracy
Measures, Shape and Space
Year R**
© A & C Black

Halves and wholes

Teachers' note Cut out the pictures and give them to the children. Ask the children to cut the pictures in half and match them up again. Encourage them to discuss each half - are they the same? As an extension, give the children a piece of paper with a clear line drawn down the middle. They choose one of the picture halves, stick it in place and then try to draw the other half.

Developing Numeracy
Measures, Shape and Space
Year R
© A & C Black

Colour the beads

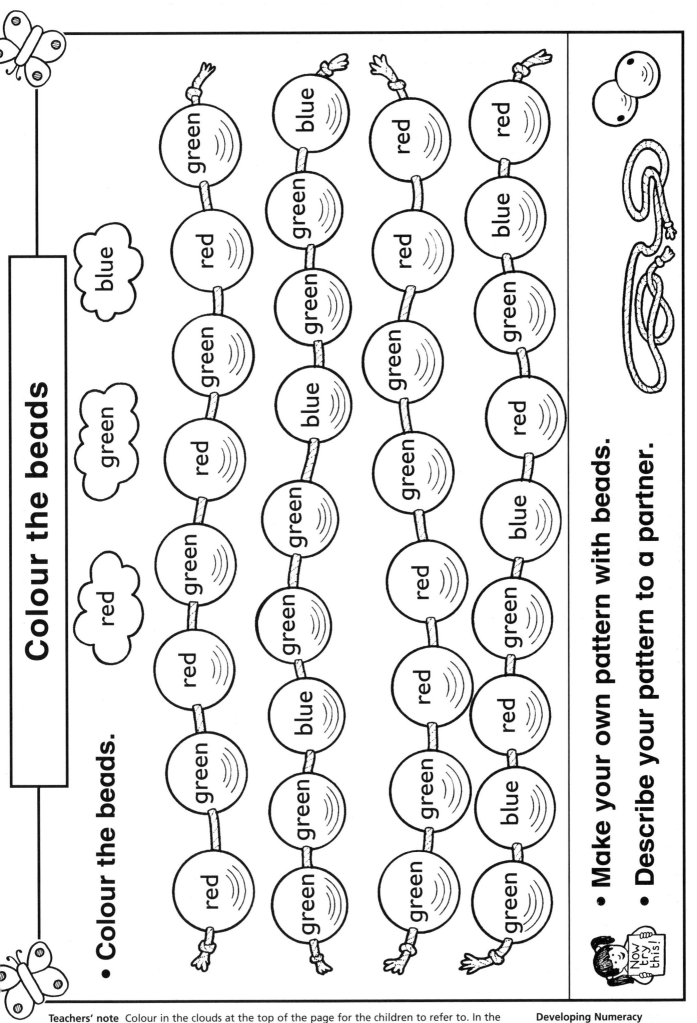

• **Colour the beads.**

red
green
blue

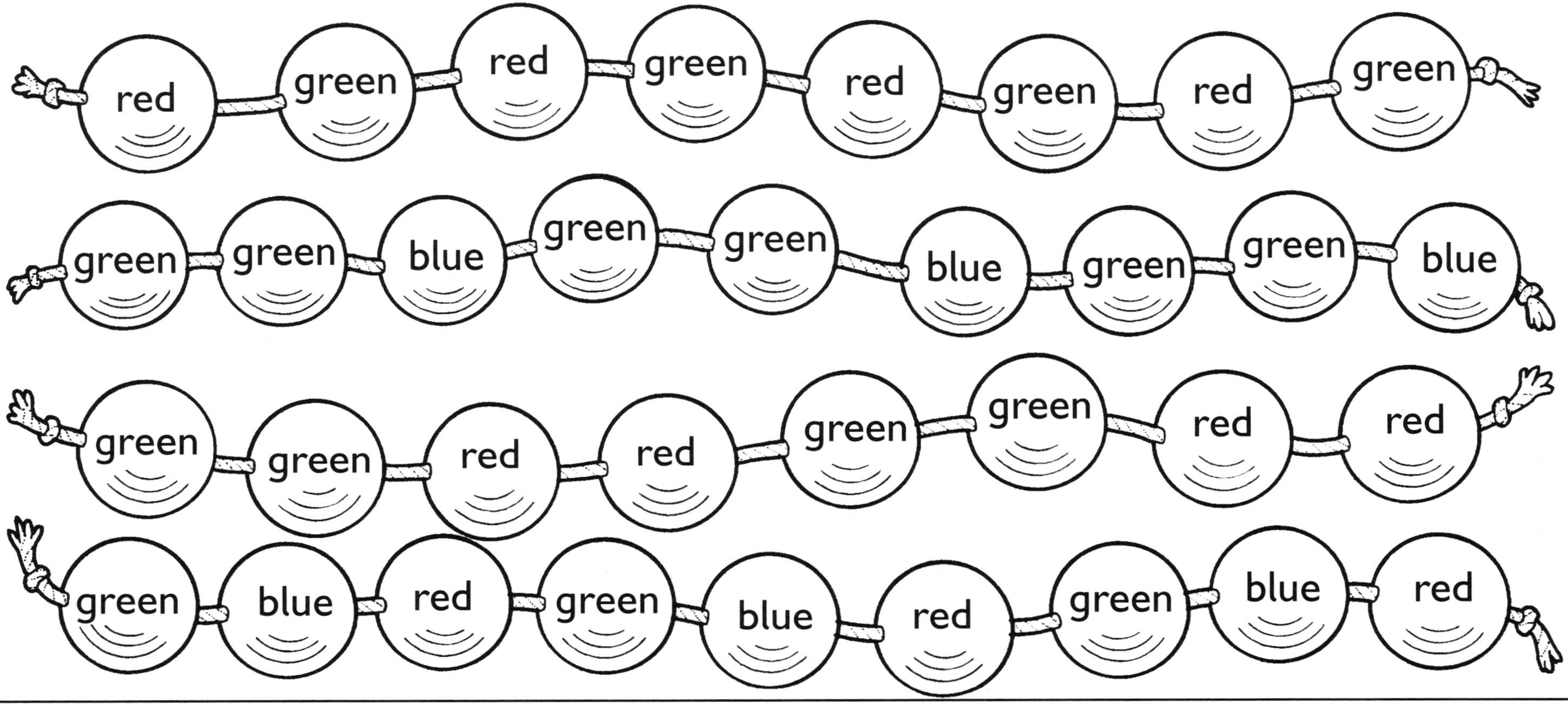
red green red green red green red green
green green blue green green blue green green blue
green green red red green green red red
green blue red green blue red green blue red

Now try this!

• **Make your own pattern with beads.**

• **Describe your pattern to a partner.**

Teachers' note Colour in the clouds at the top of the page for the children to refer to. In the extension activity, the children could use interlocking cubes if beads and laces are not available.

Bead patterns

- ## Colour the beads.

- ## Continue the pattern.

red   blue

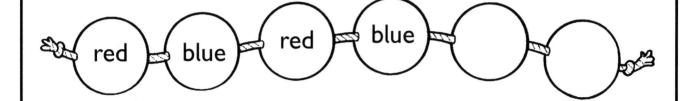

red blue red blue

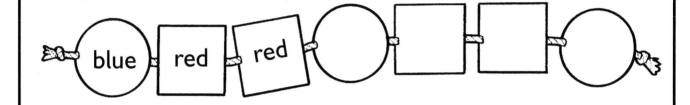

blue red red

red red blue blue

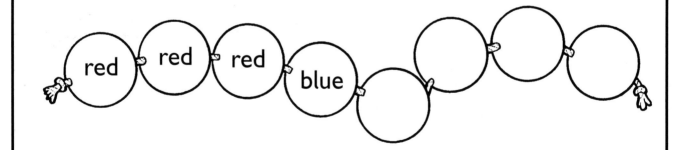

red red red blue

- ## Draw your own bead pattern.

Now try this!

Teachers' note Colour in the clouds at the top of the page for the children to refer to. If possible, provide opportunities for the children to copy the patterns with beads. Ask: 'What comes before/after the red bead?' 'What will come next?' Ensure that the children realise that shape as well as colour can be part of the pattern.

Developing Numeracy
Measures, Shape and Space
Year R
© A & C Black

Fruity patterns

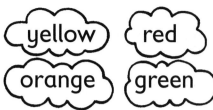

yellow red

orange green

- **Colour the fruit.**
- **Continue the pattern.**

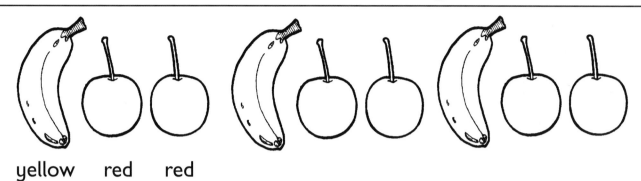

yellow   red   red

orange green yellow

red   orange   yellow

- **Colour the fruit.**
- **Continue the fruit pattern.**

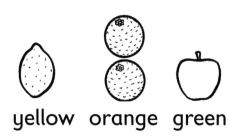

yellow orange green

Teachers' note Colour in the clouds at the top of the page for the children to refer to. Encourage the children to 'say' the patterns. Draw one of the patterns with an element missing and encourage them to say what is missing.

**Developing Numeracy
Measures, Shape and Space
Year R
© A & C Black**

Necklaces

- **Colour each necklace.**
- **Use different colours to make a repeating pattern.**

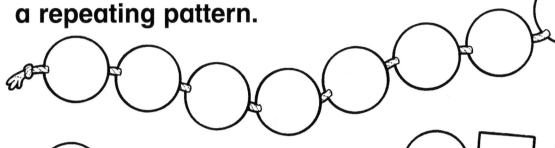

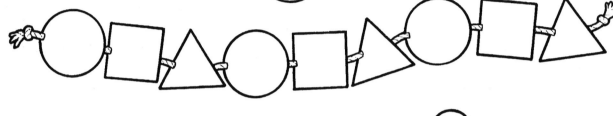

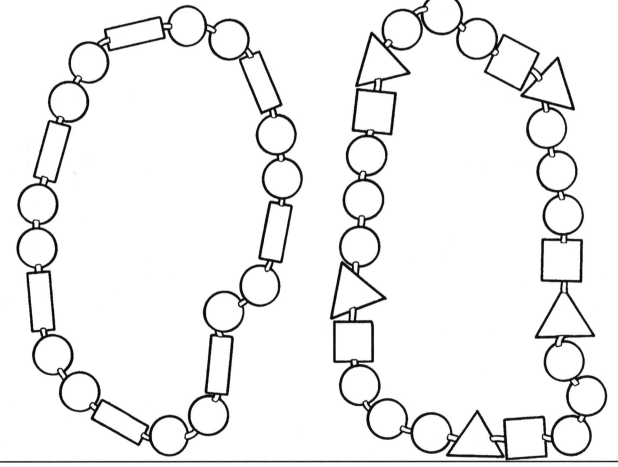

- **Draw your own necklace.**
- **Colour it to make a pattern.**

Teachers' note Encourage the children to limit the range of colours that they use for colouring the patterns. Ask them to 'say' the patterns. Where the patterns are cyclical, point out how the pattern joins up to continue around the circle. Ensure that the children realise that shape as well as colour can be part of the pattern.

Developing Numeracy
Measures, Shape and Space
Year R
© A & C Black

Teddy bears' picnic

• **Colour the teddies to make a pattern.**

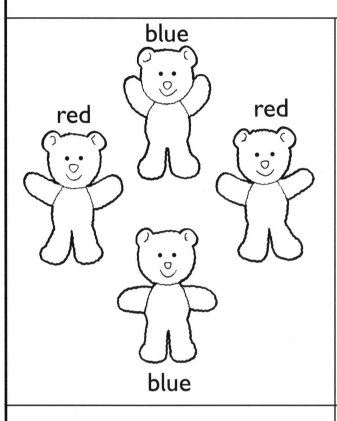

blue

red   red

blue

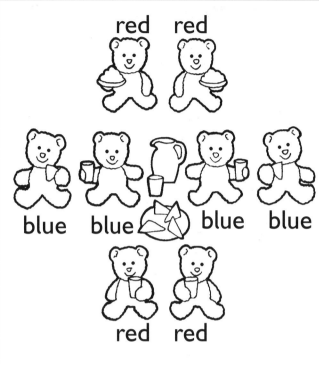

red   red

blue   blue   blue   blue

red   red

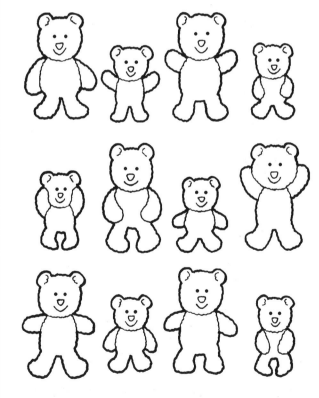

 • **Draw your own teddy pattern.**

Teachers' note Discuss the patterns and ask the children to describe them. The children could use
Compare Bears if available. Encourage the children to note the differences in size in some of
these patterns.

**Developing Numeracy
Measures, Shape and Space
Year R**
© A & C Black

In the playground

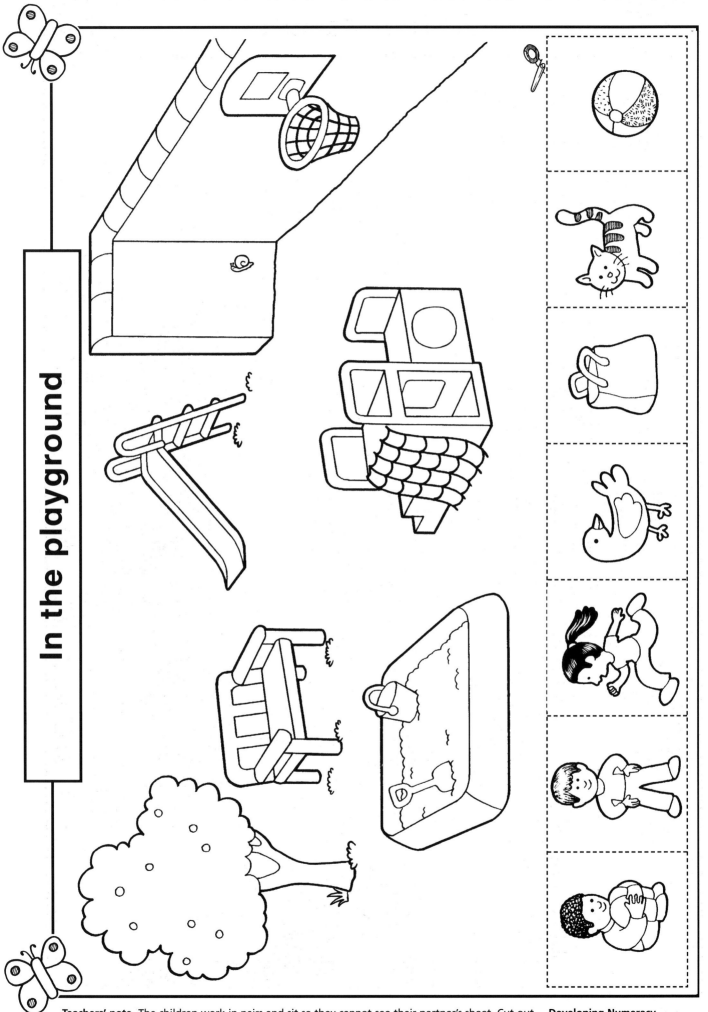

Teachers' note The children work in pairs and sit so they cannot see their partner's sheet. Cut out the objects and give them to each child. The children take turns to place an object somewhere in the playground and describe the object and its position to their partner, who then duplicates the action. When they have finished, the children compare sheets.

**Developing Numeracy
Measures, Shape and Space
Year R**
© A & C Black

At the station

- **Cut out the pieces.**
- **Make the picture.**

Teachers' note The children should work in pairs or small groups to discuss where to place the puzzle pieces. Ask them to describe where the pieces go using words such as 'next to', 'beside', 'above' and 'below'.

Developing Numeracy
Measures, Shape and Space
Year R
© A & C Black

59

Home sweet home

• **Draw the way home.**

• **Draw the way for Rabbit to visit Bird's nest.**

Teachers' note Encourage the children, working in pairs, to use their own words to describe the routes and to begin to explore the language of movement and direction. They may find it helpful to use a different-coloured crayon for each route.

Developing Numeracy
Measures, Shape and Space
Year R
© A & C Black

Little Red Riding Hood

• **Draw different ways to Grandma's house.**

• **How many different ways did you find?** ☐

Teachers' note Encourage the children, working in pairs, to find as many routes as they can. They may find it helpful to use a different-coloured crayon for each route. Ask them to describe their routes using the language of movement and direction, for example: along, past, around, turn, left, right.

**Developing Numeracy
Measures, Shape and Space
Year R**
© A & C Black

Turtle routes

• **Draw different ways for the turtle to go home.**

home					
					start

• **Draw the turtle on a different square.**

• **Find its way home.**

Teachers' note Encourage the children, working in pairs, to describe the routes using the language of movement and direction, for example 'up', 'down', 'left', 'right', and the number of squares to move. They may find it helpful to use a different-coloured crayon for each route. Ensure that they realise that they cannot move onto an 'occupied' square.

Developing Numeracy
Measures, Shape and Space
Year R
© A & C Black

Turning

• **Ring the things that** turn .

• **Draw something in the classroom**

that turns.

Teachers' note Discuss how these objects turn. It may help the children to turn the pages of a book, turn a door handle, etc. so that they experience the turning for themselves.

Developing Numeracy
Measures, Shape and Space
Year R
© A & C Black

Frog hops

Teachers' note This game is like Snakes and Ladders. The children take turns to roll a die and move their counter. If they land on a direction arrow they jump one 'pad' in that direction, for example if they land on 3 they jump to 12. Encourage them to use the language of movement and direction, and the number of lilypads. As an extension, add more arrows.

Developing Numeracy
Measures, Shape and Space
Year R
© A & C Black

64